DONE in the SUN

SOLAR PROJECTS FOR CHILDREN

By ANNE HILLERMAN

Pictures By MINA YAMASHITA

Rose

Sunstone Press
Santa Fe, New Mexico

TABLE OF CONTENTS

3 / Foreword

4 / Introduction

6 / Sun Projections

8 / Sun and Shade

10 / Sun Tattoo

12 / Sun Traps

14 / Sun Treats

16 / Sun Signs

18 / Sun Tea

20 / Sun Crystals

22 / Sunny Sundial

24 / Sun Dog Cooker

26 / Water from the Sun

29 / About the Author

31 / Bibliography

FIRST EDITION. Printed in the United States of America.

Library of Congress Cataloging in Publication Data:

Hillerman, Anne, 1949-
    Done in the sun.

    Bibliography: p. 31
    Summary: An introduction to the sun as a renewable energy source, demonstrating through simple experiements and craft projects how the sun's light and heat can be used to help us in our everyday lives.
    1. Solar energy—Juvenile literature. [1. Solar energy—Experiments. 2. Sun. 3. Experiments. 4. Handicraft]
I. Yamashita, Mina, ill. II. Title.
TJ810.H56   1983   621.47'1   83-638
ISBN: 0-86534-018-8

Published in 1983 by SUNSTONE PRESS / Post Office Box 2321 / Santa Fe, New Mexico 87501 / USA

FOREWORD

If you stop to think about it, the sun is our primary energy source, and the only source of all our renewable energy. The sun gives heat and light; it causes the wind to blow, the ocean currents to move, the plants to grow. Unfortunately, we don't often stop to think about it. The only time we pay any attention to the sun is when we study it scientifically and make it all sound terribly complicated.

The very words "solar energy" sound complex. And yet, we know all about sunrise without knowing all about astronomy. In the same way, we begin to understand solar energy long before we worry about the principles of physics or engineering.

Anne Hillerman has written a superb introduction to the sun for young people. This is a book which helps us to become familiar with our neighborhood star, the sun, and which lets us learn what the sun can do to help us in our everyday lives of work and play. By completing simple craft projects which are "Done in the Sun," we begin to understand that we can use the light and heat of the sun for our own purposes and in some unexpected ways.

Sara Balcomb
Director, American Solar Energy Society

INTRODUCTION

What if there were no sun anymore? Can you imagine what our world would be like without the sun? Think of the darkest, coldest night you can remember; there would be day after day like that! The plants we have, the animals we love, even ourselves – nothing and none of us could survive.

The sun is 93 million miles away from us but it is so powerful that it makes life possible on our planet. The sun gives us warmth and light. Plants use light from the sun in order to grow.

Many people use sun energy or "solar energy" to keep their houses and offices warm. Some schools, hospitals and other large buildings are heated with solar energy, especially in those parts of the country where the sun shines almost every day.

The sun is also used to heat water for baths and washing dishes and to make swimming pools more comfortable for swimmers.

Solar energy gets its power directly from the sun. With specially designed systems, it can also be used to make electricity.

The wonderful thing about the sun is that no matter how much of the sun's energy we use, we can never use it all up.

DONE IN THE SUN is a book about the sun and some of the many things the sun does for us. You will have a chance to put the sun's power to work for you in these projects. Along with Kevin and Peggy, you can follow Sarah's directions for a fun time in the sun. Just make sure you have all the things you'll need and read the instructions carefully before you start.

SUN PROJECTIONS

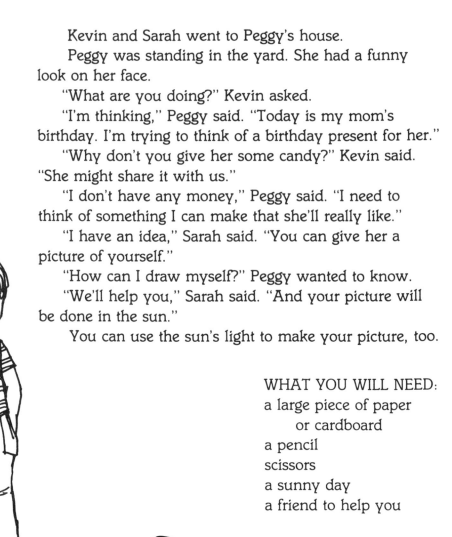

Kevin and Sarah went to Peggy's house.

Peggy was standing in the yard. She had a funny look on her face.

"What are you doing?" Kevin asked.

"I'm thinking," Peggy said. "Today is my mom's birthday. I'm trying to think of a birthday present for her."

"Why don't you give her some candy?" Kevin said. "She might share it with us."

"I don't have any money," Peggy said. "I need to think of something I can make that she'll really like."

"I have an idea," Sarah said. "You can give her a picture of yourself."

"How can I draw myself?" Peggy wanted to know.

"We'll help you," Sarah said. "And your picture will be done in the sun."

You can use the sun's light to make your picture, too.

WHAT YOU WILL NEED:
a large piece of paper
    or cardboard
a pencil
scissors
a sunny day
a friend to help you

6

WHAT TO DO:

Put the paper on the ground in the sun. It is easiest to do this project about noon when the sun is high in the sky.

Lie down on your side so you can feel the warmth of the sun on the side of your face.

Have your friend help you adjust your head and fix the paper for your shadow. The shadow of the side of your face should fall on the paper. Ask your friend to help you adjust yourself so your lips and nose will be clear.

Lie still and have your friend trace your shadow on the paper using the pencil.

Cut out the tracing. Now you have a picture of the shadow of your face. You can color it if you want to.

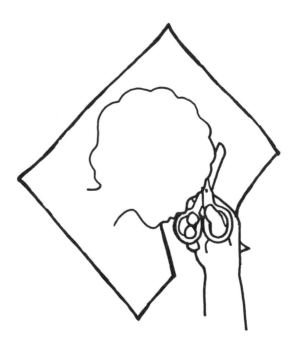

How does the sun make shadows?
Do you always have a shadow?
Is your shadow always the same size?
Can the moon make shadows?
Can a light make shadows?

SUN & SHADE

Sarah and Peggy went to Kevin's house. The three children were going to the park together.

"Wait," Kevin said. "I have to give my dog new water."

"What's wrong with the water in his pan?" Peggy asked.

"He won't drink it," Kevin said. "He doesn't like old water."

Sarah went over to the water dish.

"This water is hot," Sarah said. "That's why your dog doesn't like it."

"How did the water get hot?" Kevin asked. "I put cold water in the pan. And how can I keep it cool for him?"

"Let's do an experiment,' Sarah said. "We'll do a project about sun and shade. That will answer your questions."

8

WHAT YOU WILL NEED:
two silver-colored cans the same size
water
a thermometer (optional)

a clock or timer
a sunny day

WHAT TO DO:
Fill each can half full of cold water.
If you have a thermometer, measure
the temperature and write it down.
If you don't have a thermometer,
stick your finger in each can to be sure
the water is the same temperature.

Put one can in a spot where the sun
will shine on it.
Put the other can in the shade.

After 20 minutes, measure the
temperature of the water in each can.

Which can is cooler?

How did the water in the can in the sun get hot?
Why did the water in the can in the shade stay cool?
The water you set in the sun is hottest because it can use the
sun's heat. The water in the shade can't use the sun's power for
heat. Without the sun's energy, the water stays cool.

SUN TATTOO

It was a hot day. Kevin, Peggy and Sarah were at the swimming pool. Kevin looked at Sarah's arm.

"Sarah," he said, "you have a flower on your arm."

"I know," said Sarah. "My friend the sun made it for me. It was done in the sun."

"How did the sun do that?" Peggy wanted to know. "When I'm in the sun, I get freckles."

"I get sunburned," Kevin said. "How did you get that flower on your arm?"

"It's a sun tattoo," Sarah said, "I'll show you how to make one."

You can made a sun tattoo like Sarah's.

10

WHAT YOU WILL NEED:
adhesive tape or other tape which will stick to your skin. The tattoo is easier if you use wide tape.
scissors
a sunny day

WHAT TO DO:
Cut a small piece of the tape.
Cut the tape into the shape of a heart,
a flower, a butterfly, a lightning bolt
or whatever else you like.
The shape you cut will be the shape
of your sun tattoo.

Stick the tape on your arm where the
sun will shine on your skin.

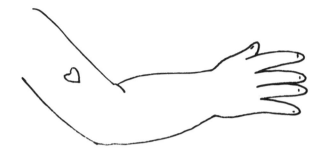

Leave the tape on all day while you
play in the sun. If the sun is bright,
you will get a good tattoo. Be careful
not to get sunburned.

Don't take the tape off early or you
won't get a tattoo.

In the evening, take off the tape.

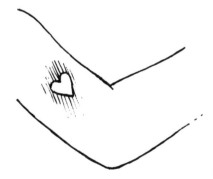

Can you see the tattoo the sun made for you? It was done
in the sun.
How did the sun make your tattoo? The sun made the rest of
your skin a little darker. That's why the tattoo can be seen.
Your tattoo will start to disappear in a day if you let the
sun shine on it.

SUN TRAPS

Kevin, Peggy and Sarah parked their bikes in the sun.

Kevin's bike has a yellow seat. Peggy's bike has a white seat and Sarah's bike has a black seat.

They saw the sun shining on their bikes.

"I bet the seat on my bike is hotter than the seats on your bikes, Sarah said.

Peggy touched the seat on her bike. It was warm. Then she touched the seat on Kevin's bike. It was hotter.

Then she touched the seat on Sarah's bike.

"Your bike is really hot," Peggy said. "It's the hottest of the three."

"How did you know your bike would be hottest?" Kevin asked.

"It's something I learned from my friend the sun," Sarah said.

"Let's do a project to see how it works."

Sarah's experiment will show you how it works.

WHAT YOU WILL NEED:
a package of all colors of construction
    paper, including black
a piece of white paper the same size
    as the construction paper
ice cubes all the same size, one for
    each different color of paper and
    one for the white sheet
small clear plastic bags
a clock or timer
a measuring cup

12

WHAT TO DO:

Take one sheet of each color construction paper and the white sheet and put them in the sun.

Put one ice cube in a bag for each sheet of paper. Act as quickly as you can so the ice won't start to melt.

Put the ice cubes in the bags on top of the paper in the sun. It's important for the sun to be shining on them.

Watch your experiment.

Which ice cubes begin to melt first?

After 20 minutes, pour the water from the bags into the measuring cup. Measure one bag of water at a time.

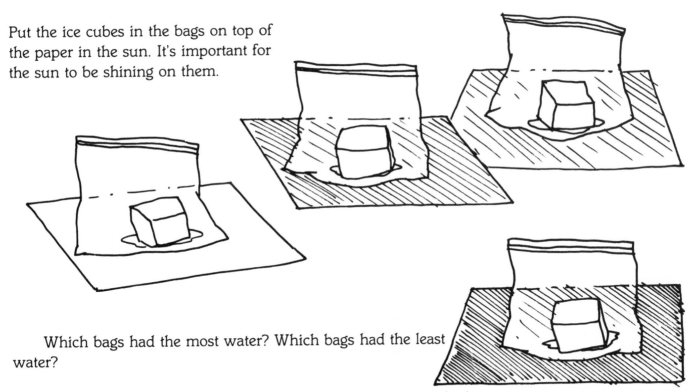

Which bags had the most water? Which bags had the least water?

The ice cubes in the bags on the darker colored papers melted fastest because the darker colors absorb the sun's heat.

The ice on the lighter colors melted more slowly because the lighter colors reflect the sun's heat. The ice cube on the white paper melted the slowest of all because white reflects the most heat.

You can do this experiment again with paper on top of the ice cubes.

13

SUN TREATS

Peggy and Kevin and Sarah were on a picnic.
"I brought the sandwiches," Peggy said.
"I brought the pickles," Kevin said.
"I brought some fruit," Sarah said.
"Wait a minute," Kevin said. "This doesn't look like fruit to me."
"Just try it," Sarah said.
"I'll try it," Peggy said, and she tasted some.
"It tastes like fruit," she said. "But why does it look different?"
"It is dried fruit," Sarah said. "It's a treat that was done in the sun. And I fixed it myself. I'll tell you how to do it."
You can make dried fruit sun treats like Sarah's.

WHAT YOU WILL NEED:
an apple, peach or apricot
a knife
a piece of cheese cloth
a paper towel
a pie pan or
    any flat glass pan or bowl
a sunny day

14

WHAT TO DO:
Wash the fruit. If you are using an apple, peel it and remove the core and seeds. If you are using a peach or an apricot, wash it and remove the seed but leave the peeling on. Cut out any brown spots, then slice the fruit into long, thin pieces. Pat the pieces dry with a paper towel.

Place the slices of fruit in the pan. Don't let them touch each other.

Cover the pan with the cheese cloth. Tuck the edges under the pan or weigh them down with rocks. The cloth is a cover for the fruit, to keep birds and insects from eating your snack.

Leave the pan in the sun during the day. Bring it inside at night.

Plan to leave your fruit in the sun for at least eight hours. If the days are hot and the sun is bright, the fruit will dry that fast. Otherwise, it may take longer.

When the fruit slices are dry, taste your snack. It was done in the sun.

Does the dried fruit look like the slices you put in the pan?
How is dried fruit different from fresh fruit?
What did the sun do to the fruit?
What are some other dried fruits you like to eat?

15

SUN SIGNS

Kevin looked sad.

"What's the matter?" Peggy asked him

"I've got a big problem," he said. "I want to make a sign for my room to put my name on the door."

"I'll help you spell," Sarah said.

"I can spell," Kevin said. "That's not the problem."

"So what's wrong?" Peggy asked.

"I don't have any crayons," Kevin said. "I don't even have a pencil."

"Well," Sarah said, "my friend the sun can help you."

So she took the scissors and cut out a K and an E and a V and an I and an N. She put the letters on a sheet of paper.

"Now we can let the sun work for us," she said.

When Peggy and Kevin came back, the sign was done. It said KEVIN. It was done in the sun.

You can make a sun sign too.

WHAT YOU WILL NEED:
construction paper
scissors
a pencil
a calm, sunny day
a clock or watch
    you may also want to use:
½ cup of dried beans
some small rocks or
    gravel would be fine

WHAT TO DO:

First, print your name or message in big block letters on the construction paper. Then cut them out.

Place the letters on a second piece of construction paper. Arrange the letters to spell your name or to form the message you want. Weigh them down with the gravel if it is breezy. If the day is too windy, save this experiment for another time.

Find a sunny spot where you can leave your project. Let the letters or message sit in the sun for three or four hours. Don't move it or the letters will not be clear.

When the time is up, remove the letters you cut out. What do you find underneath?

The longer you leave your message in the sun, the darker the letters will look. You can also make a message with rocks or beans. Or you can cut a design from paper.

How did the sun print your message? The sun's light fades the paper, except where it is shaded by the letters you cut out. That's why the letters look darker. It was done in the sun.

SUN TEA

Sarah was on her front porch with a jar of water and some peppermint tea bags. Peggy and Kevin walked up to her.

"What are you doing?" Peggy asked. "It's hot today. Let's go swimming."

"Okay," said Sarah. "And while we're at the pool, my friend the sun will make us some tea."

"Oh yeah?" asked Kevin. "Only grownups can make tea."

"You'll see," said Sarah. "The sun will do the cooking."

When Sarah and Peggy and Kevin got back from their swim, they looked in the jar where Sarah put the tea bags.

"It's magic," Kevin said. "The sun did make our tea."

"It's not magic," Peggy said. "It was done in the sun."

You can make sun tea like Sarah's.

18

WHAT YOU WILL NEED:
a quart jar with a lid
three bags of peppermint tea or
    another tea you like
water
a sunny day

WHAT TO DO:
Put the tea bags in the jar and fill
the jar with water.

Put the lid on the jar and set the
jar in the sun.

On a bright warm day you'll have
sun tea in an hour or two.
It was done in the sun.

When the tea is ready — you can tell
by the dark color — remove the tea
bags. Pour the tea into a glass with ice
cubes and taste the sun's good work.

19

How did the sun make your tea?
    The water in the jar absorbs the sun's heat. The sun makes the
water warm and the warm water cooks the tea for you.

SUN CRYSTALS

Sarah walked carefully, carrying something in her hands.

Peggy and Kevin were giving their dog a bath as Sarah came up to them with a smile on her face.

"Look what I have," she said.

Peggy and Kevin looked. Sarah showed them a pan filled with pretty white crystals.

"It looks like tiny white mountains," Kevin said.

"It looks like a forest covered with snow," Peggy said.

"It makes me think of the surface of the moon," Sarah said. "I call it my crystal garden."

"Where did you buy it?" Kevin wanted to know.

"I didn't buy it," she told him. "I made it with the help of my friend the sun. It was done in the sun."

"Show us how to make one," Peggy and Kevin said. And so Sarah did.

You can make a crystal garden like Sarah's.

WHAT YOU WILL NEED:
a large flat cake pan or
        a large pie pan
salt
a measuring cup
water
a sunny day
a magnifying glass or hand lens

WHAT TO DO:
Put about ½ cup of salt in the pan. Add a cup of water and mix to dissolve the salt as well as you can.

Set the pan in the sun. Leave it out for as long as it takes the water to disappear completely. Don't move the pan until the water is gone. It could take two or three days or more. Be patient.

When the water has dried up, you will have a garden of white crystals.

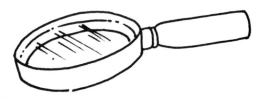

Look at the crystals through the magnifying glass. What do they look like to you? Do you have any that look like icicles? Do some look like lace? Do the crystals look like the salt you first put in?

What happened to the water in the pan?

The heat of the sun made the water disappear. This is known as evaporation.

21

SUNNY SUNDIAL

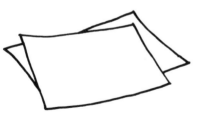

Peggy and Kevin and Sarah were playing in Sarah's yard. Peggy had to go inside to see what time it was. She had to be home at 6 o'clock.

"I wish we had a clock out here," Peggy said.

"Me, too," said Kevin.

"I'll show you how to tell time without a clock," Sarah said.

Sarah brought out a spool, a pencil, a piece of construction paper and other things.

"You can't tell time with that junk," Kevin said.

"You'll see," said Sarah. And she showed them how to make a sundial. They worked on it the next day, too.

When the sundial was finished Peggy said, "Now I can tell time without a clock."

"You don't have to wind a sundial to make it work," Kevin said. "It's done in the sun."

You can make a sundial, too.

WHAT YOU WILL NEED:
a piece of construction paper
a spool
a ruler
two pencils
tacks
a sunny day
a clock

WHAT TO DO:
Tack the construction paper to a board
or another flat surface.

Place the spool in the center of the
paper. Draw a circle around the spool
to mark the spot in case it gets
bumped out of place.

Stick a pencil inside the spool.

Watch the clock. Every hour, mark
the place on the paper where the
shadow falls. Use the ruler to make
the line straight. At the end of the line
write what time it is.

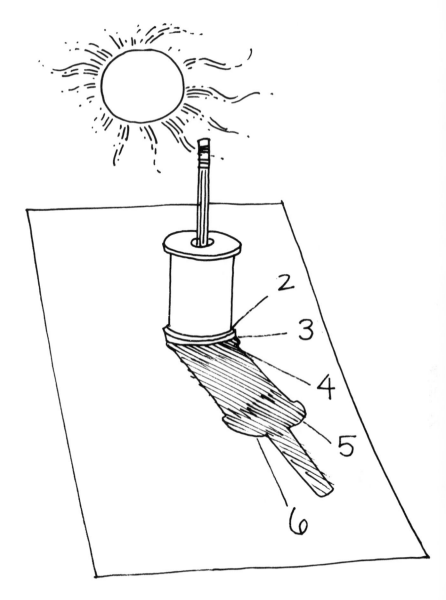

When you are done, you will be able to tell what time it is by
watching where the shadow falls on your sundial.
  Does the sundial work at night?
  Where do the shadows come from?
  Why do the shadows move?
  The shadows move because the sun is moving across the sky.
The light from the sun causes the shadow.

23

SUN DOG COOKER

Sarah was sitting on the grass eating a hot dog.

"That looks good," Kevin said. "Would your mom fix one for me?"

"I cooked it myself," Sarah said. "And it's not a hot dog. It's a sun dog."

"Will you please fix a sun dog for me?" Kevin asked.

"And for me, too," Peggy said.

"We can all cook sun dogs like this," Sarah said. "It was all done in the sun."

You can fix a hot dog the way Sarah did.

WHAT YOU WILL NEED:
an empty oatmeal box with its lid
a wire coat hanger or
    a piece of heavy wire
wire cutters
aluminum foil
scissors
tape
a hot dog
a sunny day

WHAT TO DO:

Cut out half of the oatmeal box from the side. Make it look like a funny boat.

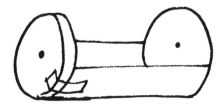

Straighten the hanger and use the wire cutters to cut a piece of it. The piece should be at least three inches longer than the oatmeal box. Or cut a piece of heavy wire.

Tape the top onto the box. Make sure it won't fall off.

Line the oatmeal box with foil. Don't let any cardboard show.

Stick the hanger or wire through the middle of the bottom of the box and through the hot dog from end to end.

Then stick the hanger or wire through the other end of the box to the outside.

Set the hot dog in the cooker in the sun. The open side of the box must face the sun for the hot dog to cook. Turn the hanger or wire so the hot dog cooks on all sides.

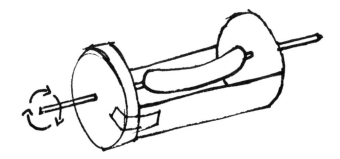

When your hot dog is cooked you can call it a sun dog. Take it out of the cooker. Your sun dog is ready to eat.

How did the sun dog cook?

Why did you use the foil?

Did the sun dog taste like a hot dog cooked over a campfire or on a stove.?

Did the sun give the sun dog a flavor?

Can you think of anything else you could cook in the sun dog cooker?

WATER FROM THE SUN

One day Sarah and her friends were talking.

"What would you do if you were stuck on an island that didn't have any water to drink?" Peggy asked.

"I'd drink from the sea," Kevin said.

"Oh, no," said Peggy. "The sea is too salty."

"I'd make my own drinking water," Sarah said. "With my friend the sun to help me, I could do it. It would be done in the sun."

"Oh, really?" said Kevin. "How could you make water to drink?

"Let's do an experiment," Sarah said. "Then you'll see."

WHAT YOU WILL NEED:
a large bowl
a small, shorter bowl
a measuring cup
salt
heavy-duty plastic wrap
a small rock or other weight
a sunny day

WHAT TO DO:
First, put ¼ cup of salt into the large bowl. Put a cup of water in the bowl and stir until the salt dissolves.

Set the large bowl in the sun.

26

Carefully put the small bowl inside the large bowl near the center. Be sure no salt water gets in the small bowl. That will spoil the experiment.

Cover the top of the large bowl with plastic wrap. Pull the plastic wrap tightly over the top and press it around the edges. If it rips, try again with another piece.

When you check on your experiment, you'll see tiny drops of water on the plastic wrap. Because of the weight, the drops will drip into the small bowl.

When you have some water in the small bowl, remove the plastic and taste the water. The longer you leave the bowls in the sun, the more water you will get in the small bowl.

Place the rock on the plastic over the center of the small bowl.

Set the bowls in the sun for two hours.

What does the water in the small bowl taste like?
Now taste the water in the big bowl. Which water is saltier?
Is the water warm?
Where did the heat come from?
The water in the small bowl may taste a little salty. It will be less salty than the water in the big bowl.
Evaporation caused the water to drip into the small bowl. The salt was left behind.

27

ABOUT THE AUTHOR

ANNE HILLERMAN was born in 1949 in Lawton, Oklahoma, and moved to Santa Fe with her parents in 1953. The eldest of six children, Anne grew up in Santa Fe and Albuquerque in a family that was very interested in reading and writing. She attended the University of New Mexico and the University of Massachusetts and graduated from UNM with a degree in journalism. She has worked as a legislative correspondent, education writer, feature writer, radio and television newscaster and writer, performing arts reporter and critic and editor of an arts and entertainment magazine.

Currently Anne is editorial writer for The New Mexican, Santa Fe's daily newspaper, and an officer in New Mexico Press Women association. She has won numerous regional and national awards for her writing.

She and her father, author Tony Hillerman, collaborated on several editions of Fodor's Travel Guides. She has also written travel and adventure articles for magazines as a freelance writer.

Anne and her husband live in Santa Fe where they have built three homes which utilize solar features. She enjoys skiing, hiking, classical music, reading and working with the Santa Fe Girls' Club.

BIBLIOGRAPHY

Adams, Florence. Catch a Sunbeam: A Book of Solar Study & Experiments. New York: Harcourt Brace Jovanovich, 1978. **

Andrassy, Stella. The Solar Cookbook. Dobbs Ferry, NY: Morgan, 1981.

Asimov, Isaac. How Did We Find Out About Solar Power? New York: Walker, 1981. **

Baer, Steve. Sunspots: An Exploration of Solar Energy through Fact & Fiction. (rev.ed.) Point Roberts, WA, Cloudburst, 1979. **

Behrman, Daniel. Solar Energy: The Awakening Science. Boston: Little, Brown, 1980.

Bendick, Jeanne. Putting the Sun to Work. Westport, CT: Garrard, 1978. **

Berger, Melvin. Energy from the Sun. New York: Harper & Row, 1976. **

Brinkworth, Brian Joseph. Solar Energy for Man. New York: Wiley, 1972.

Catherall, Ed. Solar Power. Morristown, NJ: Silver Burdett, 1981. **

Farrands, Barry J. Everything You Always Wanted To Know About Solar Energy, But Didn't Know Who To Ask. Wilmington, DE: Promise Corp., 1980.

Gadler, Steve, & Adamson, Wendy. Sun Power: Facts About Solar Energy. (rev.ed.) Minneapolis: Lerner, 1979. **

Hoke, John. Solar Energy. (rev.ed.) New York: Watts, 1978. **

Kaplan, Sheila. Solar Energy. Milwaukee, WI: Raintree, 1982. **

Knight, David. Harnessing the Sun: The Story of Solar Energy. New York: Morrow, 1976. **

McDaniels, David K. The Sun: Our Future Energy Source. New York: Wiley, 1980.

Metos, Thomas H., & Bitter, Gary G. Exploring With Solar Energy. New York: Messner, 1978. **

Smith, Norman. Sunpower. New York: Coward, McCann & Geohegan, 1976. **

Smithsonian Institution. Fire of Life: The Smithsonian Book of the Sun. Washington, DC: Smithsonian, 1981.

Spetgany, Tilly, & Wells, Malcolm. The Children's Solar Energy Book: Even Grown-Ups Can Understand. New York: Sterling, 1982. **

Spooner, Maggie. Sunpower Experiments: Solar Energy Explained. New York: Sterling, 1979. **

Washburn, Mark. In the Light of the Sun: From Sunspots to Solar Energy. New York: Harcourt Brace Jovanovich, 1981.

Yates, Madeleine. Sun Power: The Story of Solar Energy. Nashville, TN: Abingdon, 1982. **

31

** Books for children.

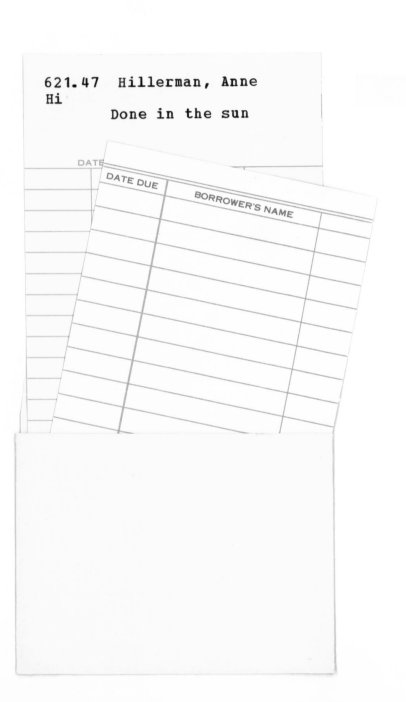